The Tiny Astronaut

A Solar System Adventure for Little Explorers

by
R. J. Hudson
The Tiny Books Group™

The Tiny Books Group™

thetinybooksgroup.com

Copyright © 2023 R. J. Hudson

All rights reserved.

The stars above are burning bright,
Our Solar System waits, to share cosmic delights.

Come with me, Tiny Astronaut friend,
I will show you the place where planets and stars never end...

THE SUN

First stop, The Sun, our fierce and fiery star.
In the middle of the solar system, with the most important job, by far.

With its golden glow, it warms our day,
And in the night, it hides away.

MERCURY

Mercury, the first planet, is closest to the Sun.
The smallest in the solar system and fastest in its orbit, bar none.

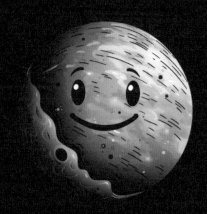

With giant craters and rocks on its surface, beware.
Its thin atmosphere means that you cannot breathe air.

VENUS
2

The second planet is Venus, with her thick, cloudy shroud,
She's a blazing furnace, where nothing is allowed.

The hottest planet, and WOW what a sight,
With her glowing clouds of yellowish light.

3 EARTH

Say Hello to Earth; our home. She's planet number three,
Her blue and green beauty is easy to see.

With oceans and mountains, filled with wonders galore.
Endless opportunities are there for you to explore.

Mars, the "Red Planet," someday we might visit,
With its deserts and canyons, you wouldn't want to miss it.

No humans have set foot on the fourth planet before,
but 5 robots called "rovers" have all landed ashore.

5
JUPITER

Jupiter is the giant, so massive and vast,
with swirling storms that forever last.

Planet five has 80 moons,
short days and huge rings. Its name is fit for a Roman God King.

6
SATURN

Saturn's always smiling, showing off its rings of ice.
The most moons of any planet, 124 to be precise!

Number six is quite impressive, tell me if you agree.
On a clear dark night look into the sky and you'll see.

7 URANUS

Number seven is Uranus, a planet ice cold and blue. The only one tilted on its side, a fascinating view. "Ice Giant", the chilliest world of all, Four times Earth's size, a magnificent frozen ball

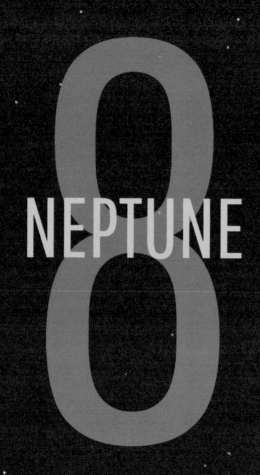

Last but not least here is Neptune; "Big Blue",
Roman god of the sea and he's smiling at you!

With winds that howl all night and all day,
3 billion miles from Earth and twins with Uranus, some say.

Now, Tiny Astronaut our journey is done.
Around the universe and back again, and twice around The Sun.

Sleep tight, little one, in your bed; cozy and soft. Your adventures are launching...

3, 2, 1, blast off!

by
R. J. Hudson
The Tiny Books Group
 @thetiny__books
tinybookgroup@gmail.com

A little note from the author:
Thank you for purchasing my book, if you and your little ones enjoyed it, please take a few moments to leave an honest review on Amazon.

Scan the QR code to leave a review

The Tiny Books Series

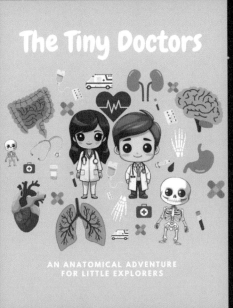

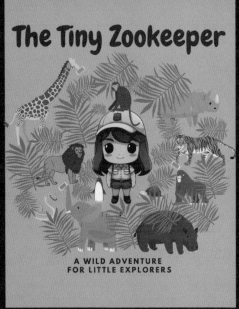

Follow 📷 @thetiny__books

Subscribe at: tinybookgroup@gmail.com

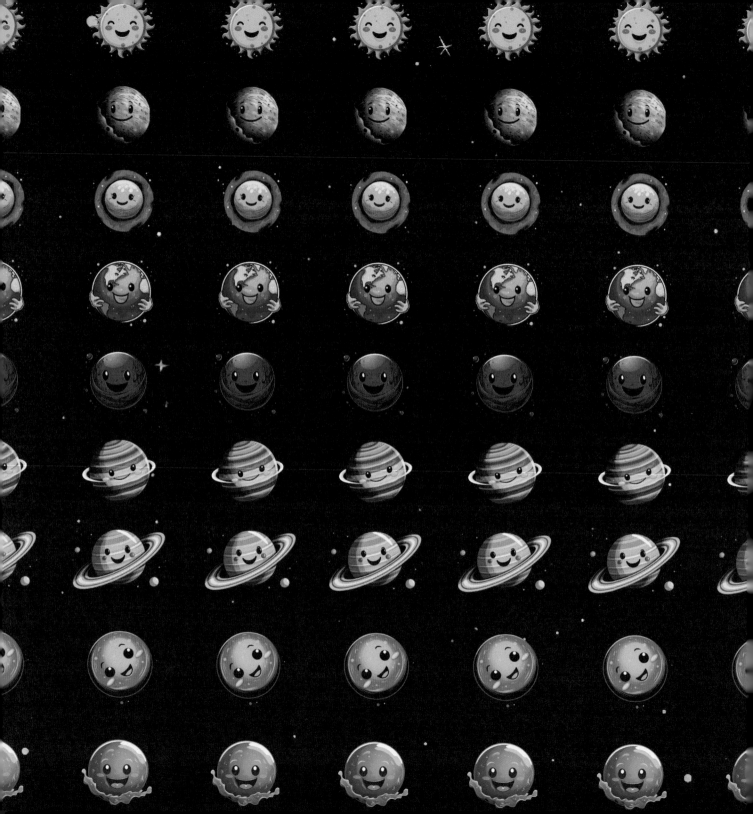

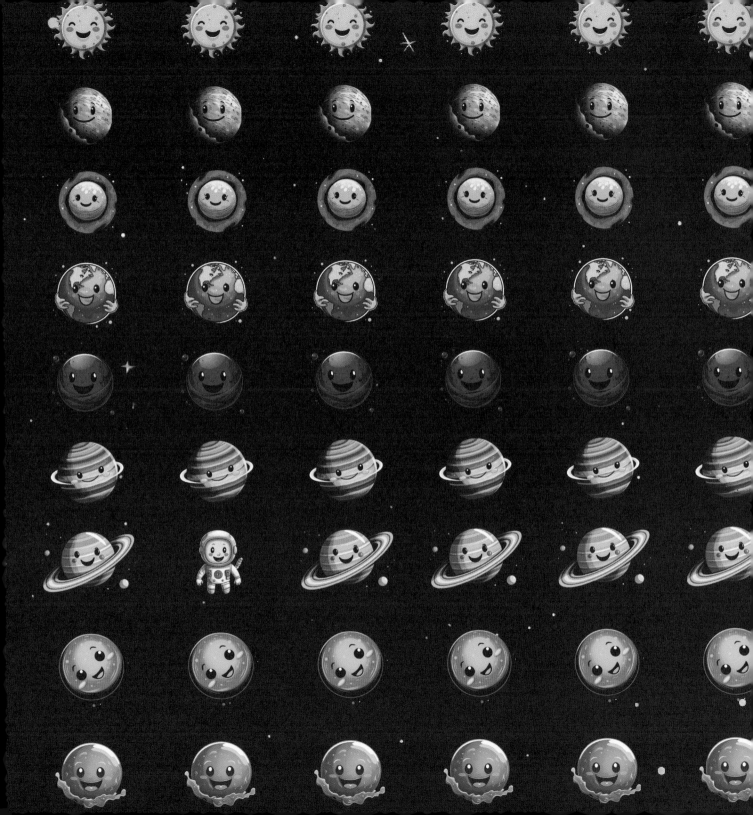

Printed in Great Britain
by Amazon